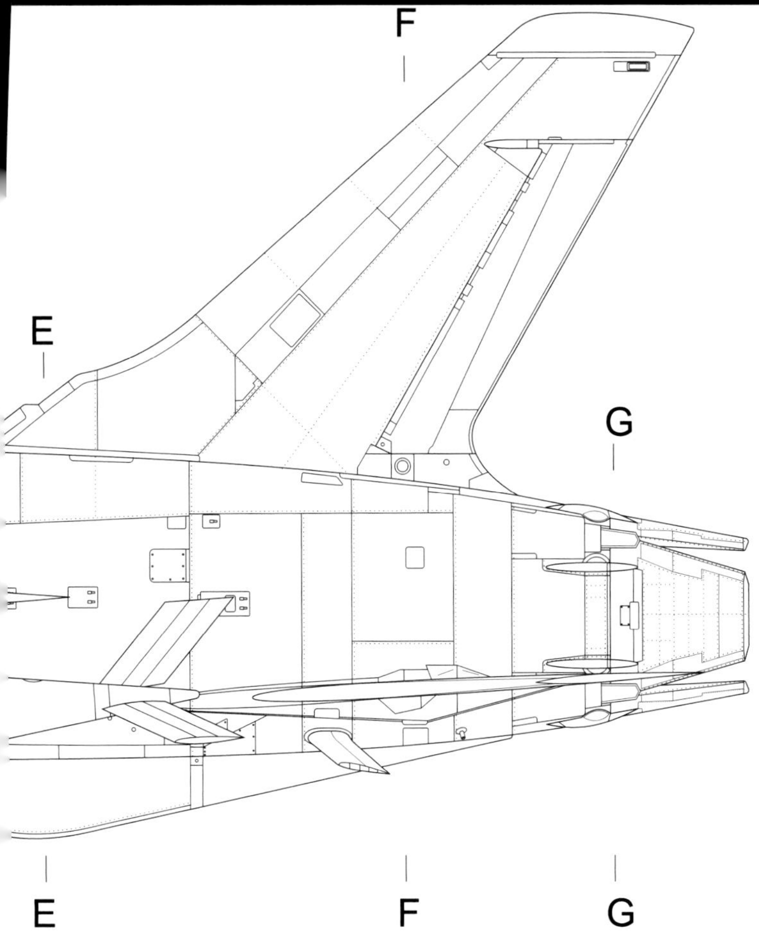
F
E
G
E
F
G

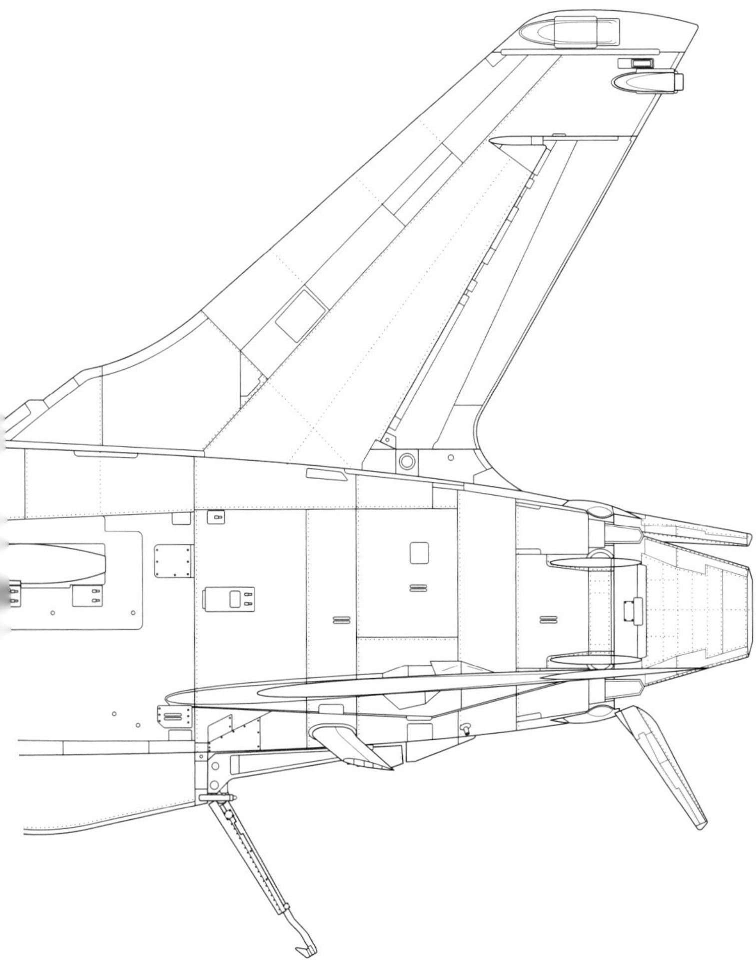

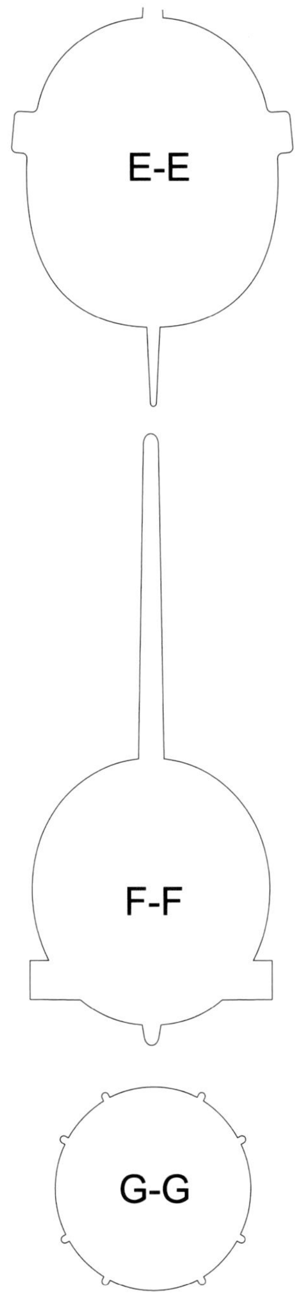
E-E
F-F
G-G

F-105D-31-RE

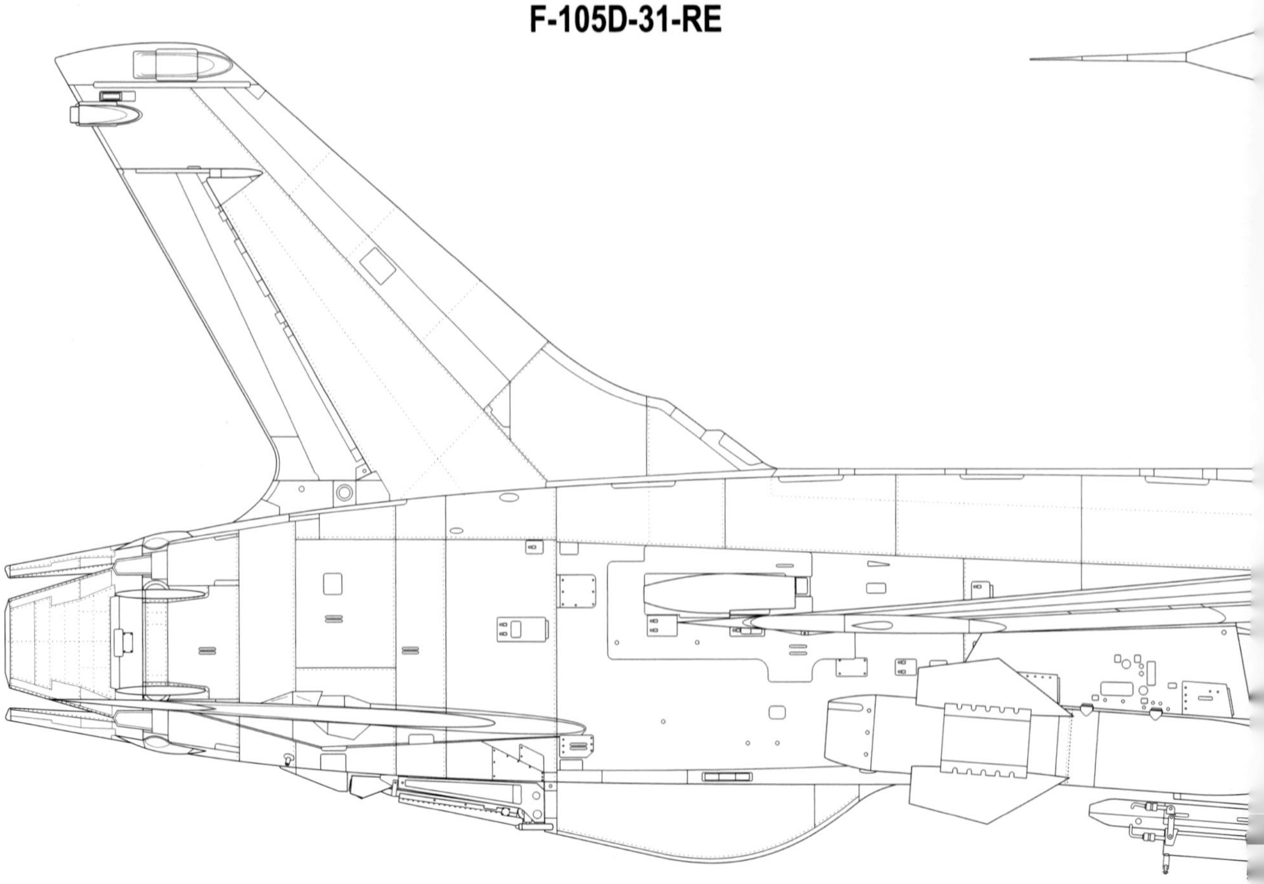

F-10

with A
on inb

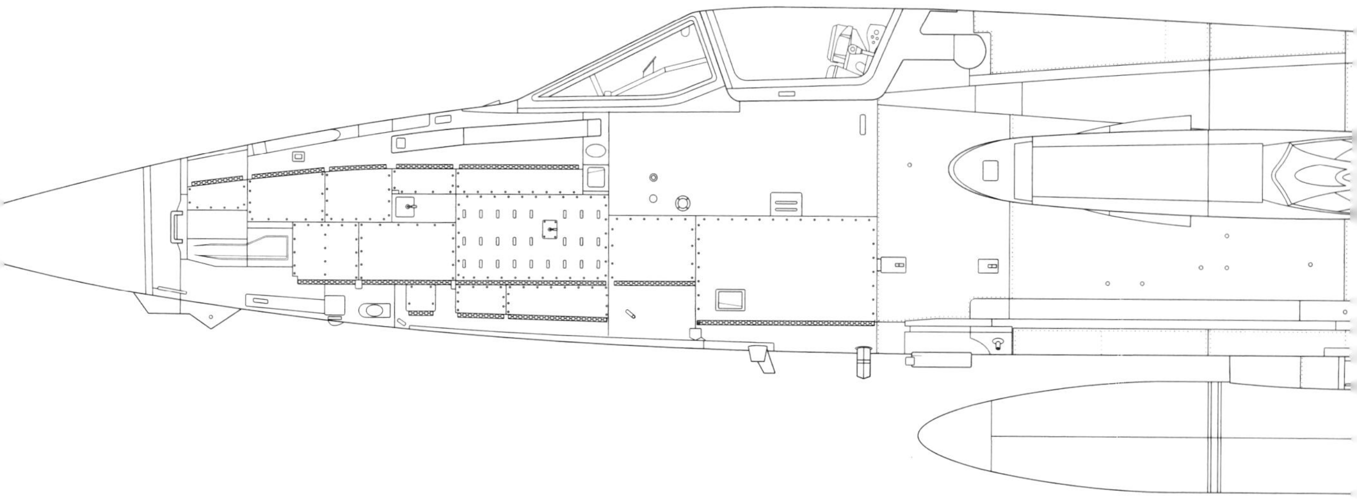

2.460 liters fuel tank

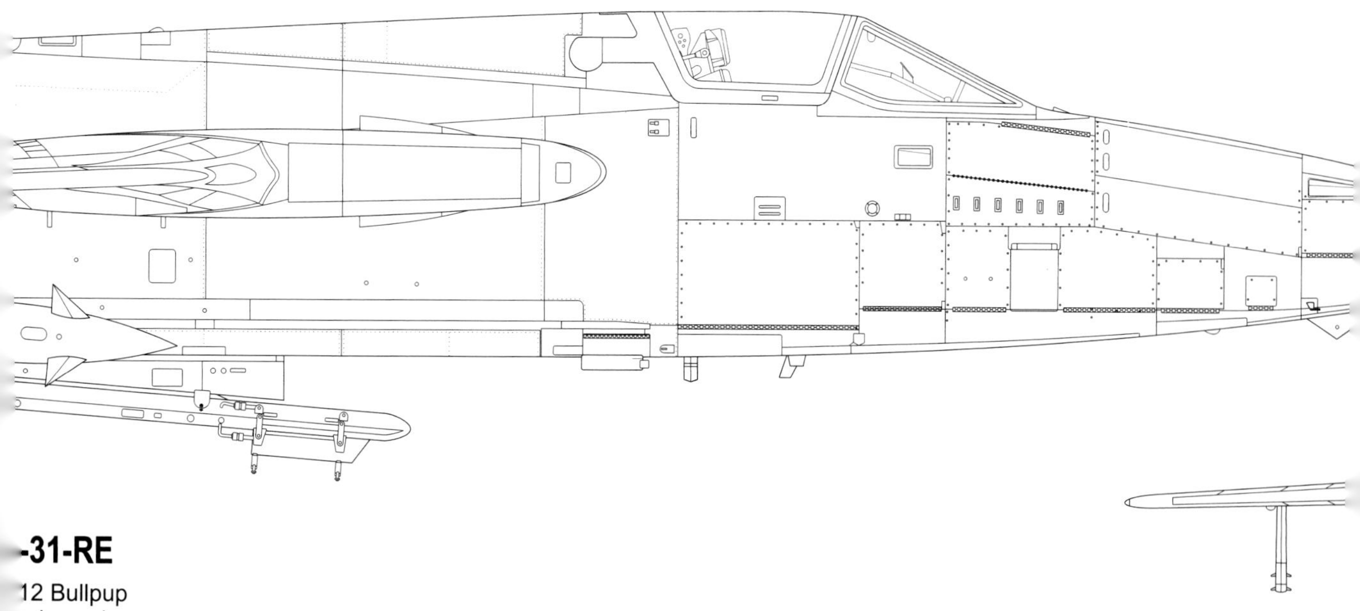

-31-RE

12 Bullpup
wing pylon

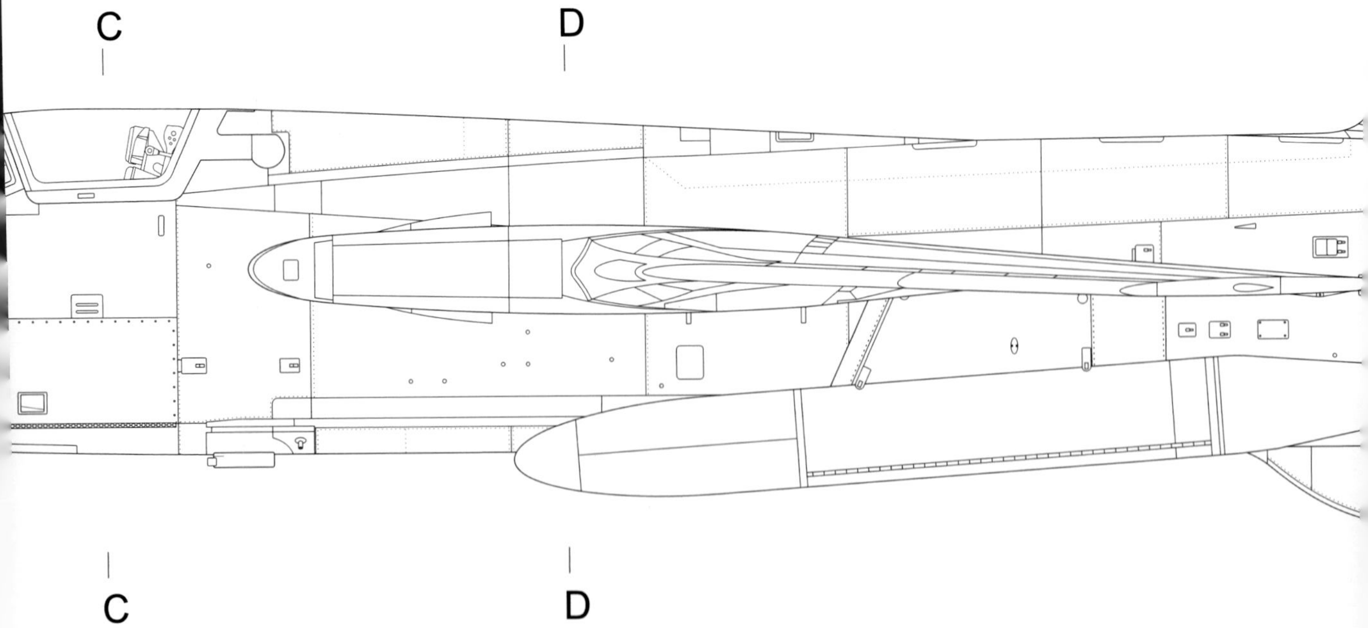

REPUBLIC F-105 THUNDERCHIEF

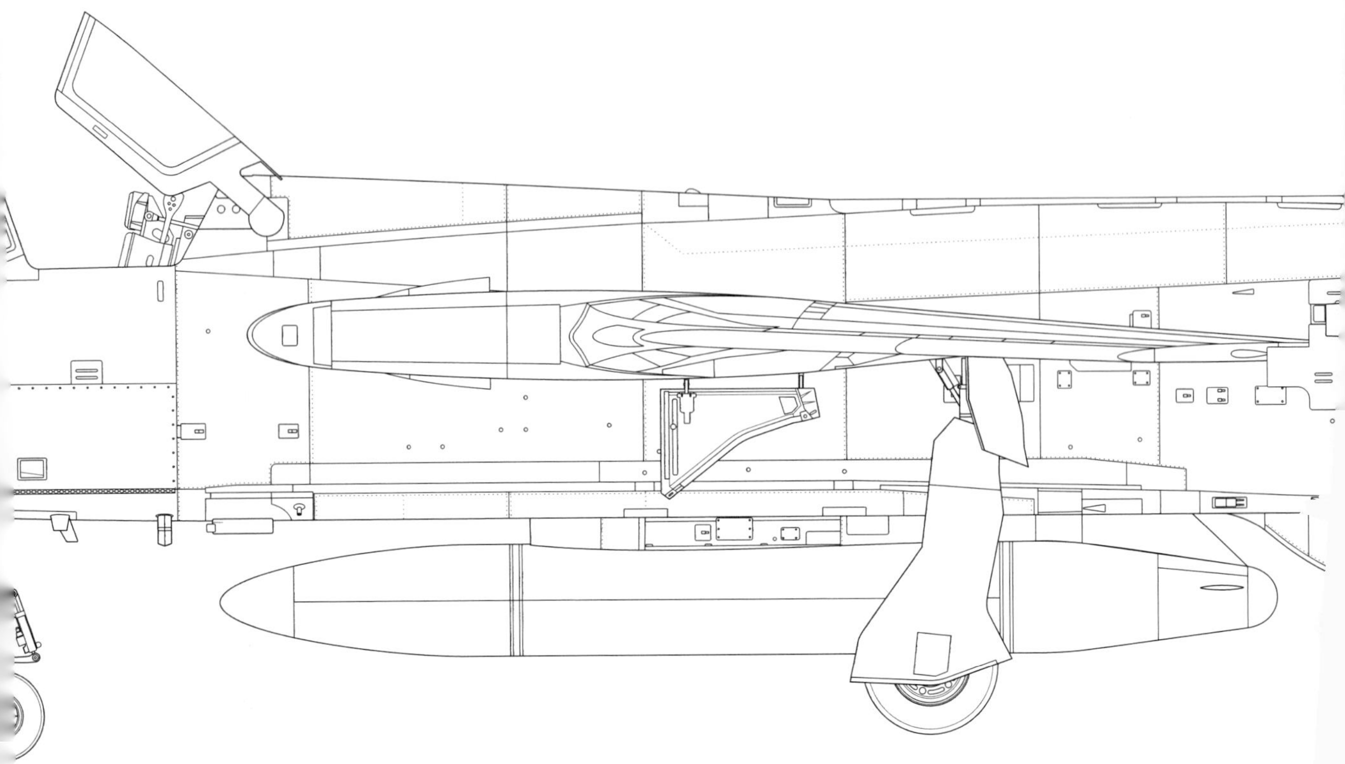

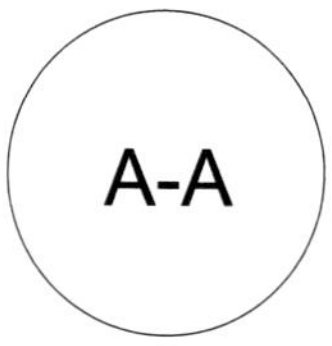

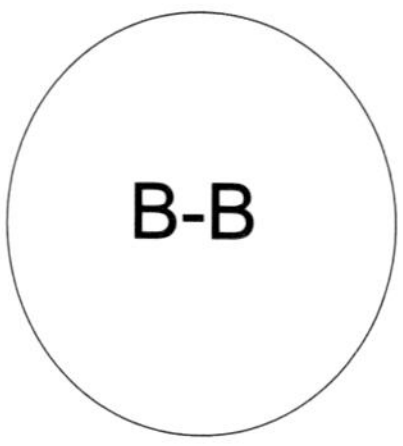

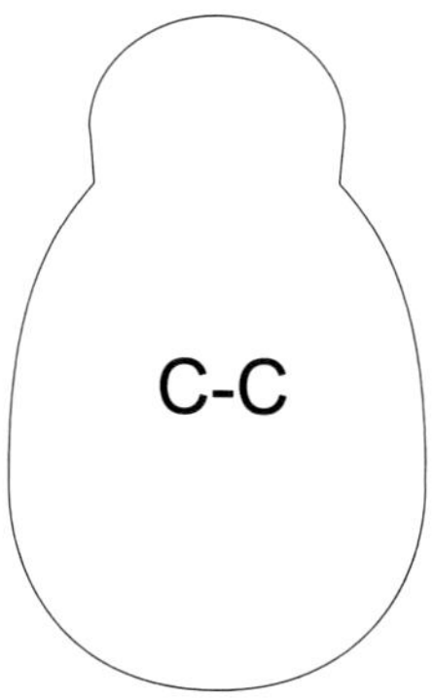

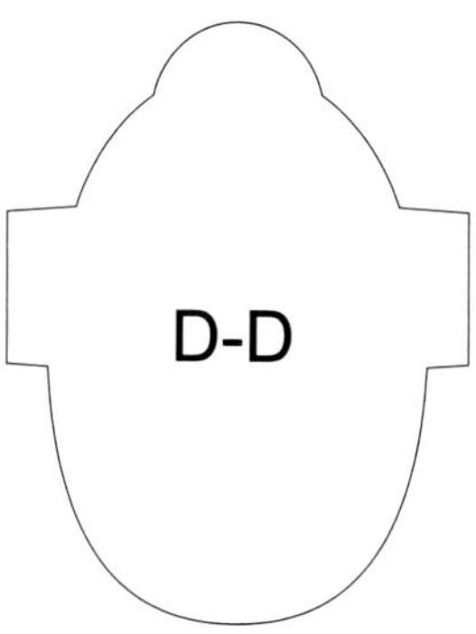

F-105D-10-RE
with 1.703 liters fuel tank
on inboard pylon

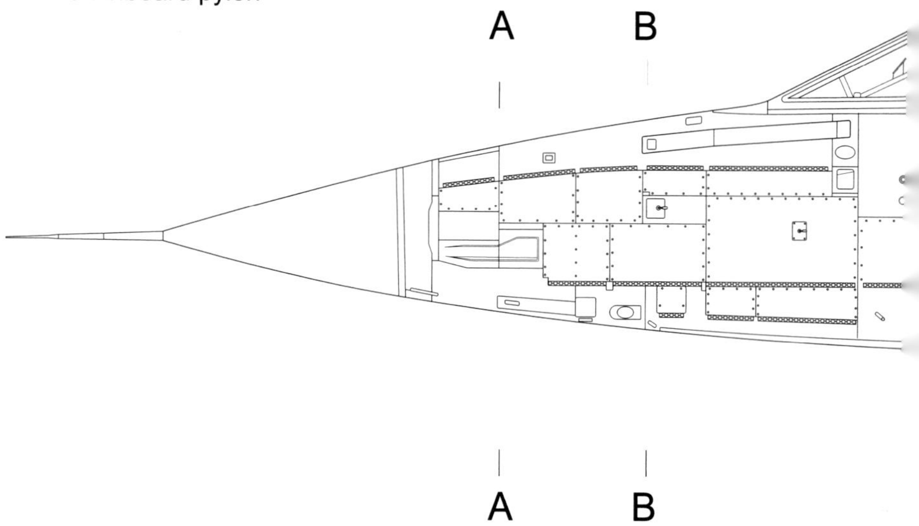

F-105D-31-RE

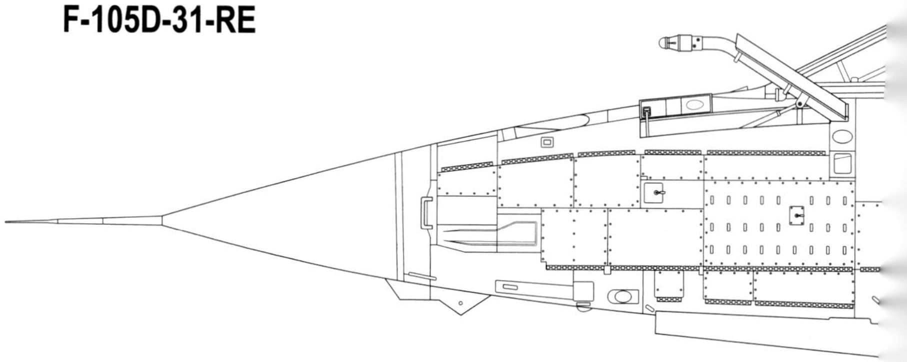

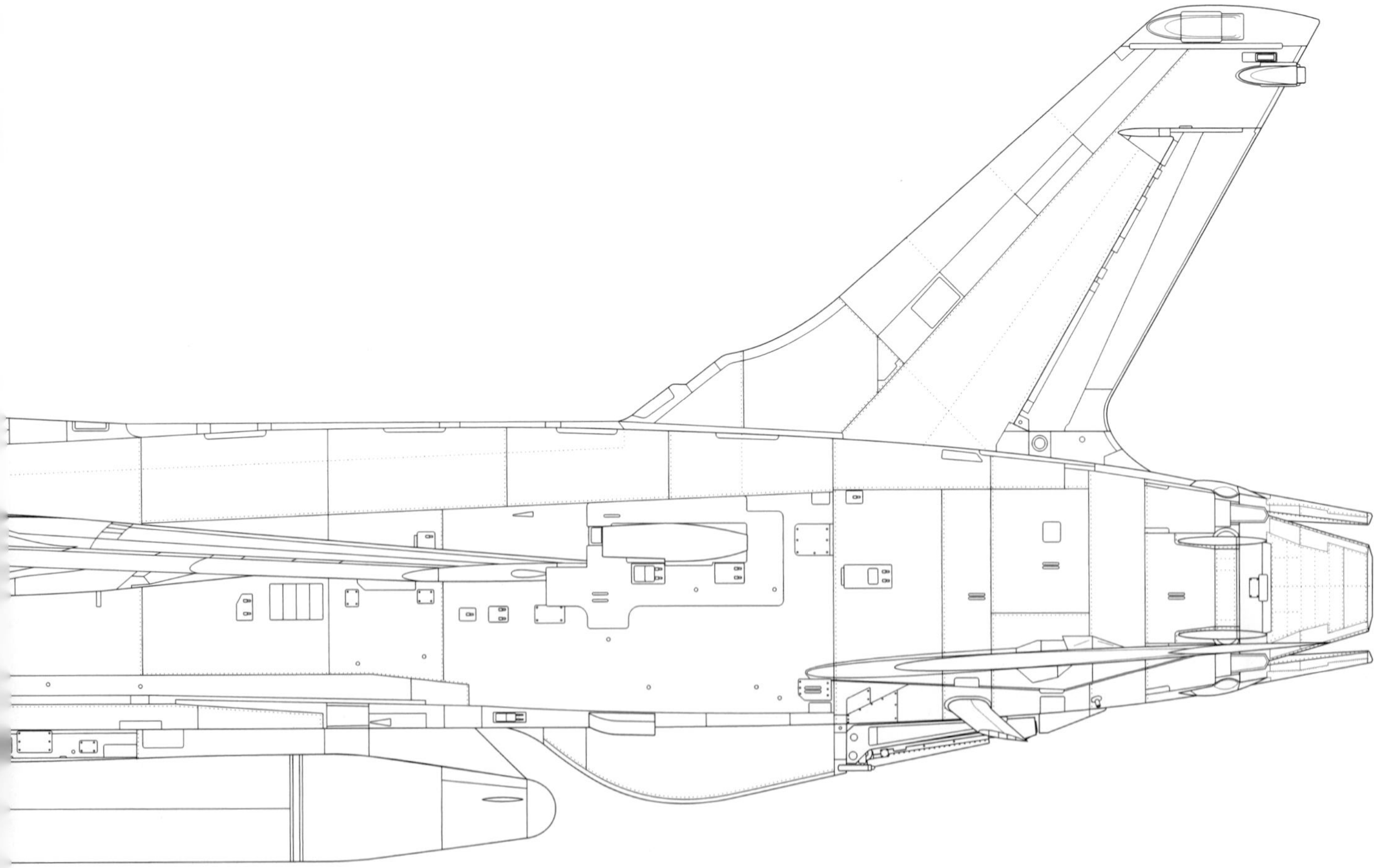

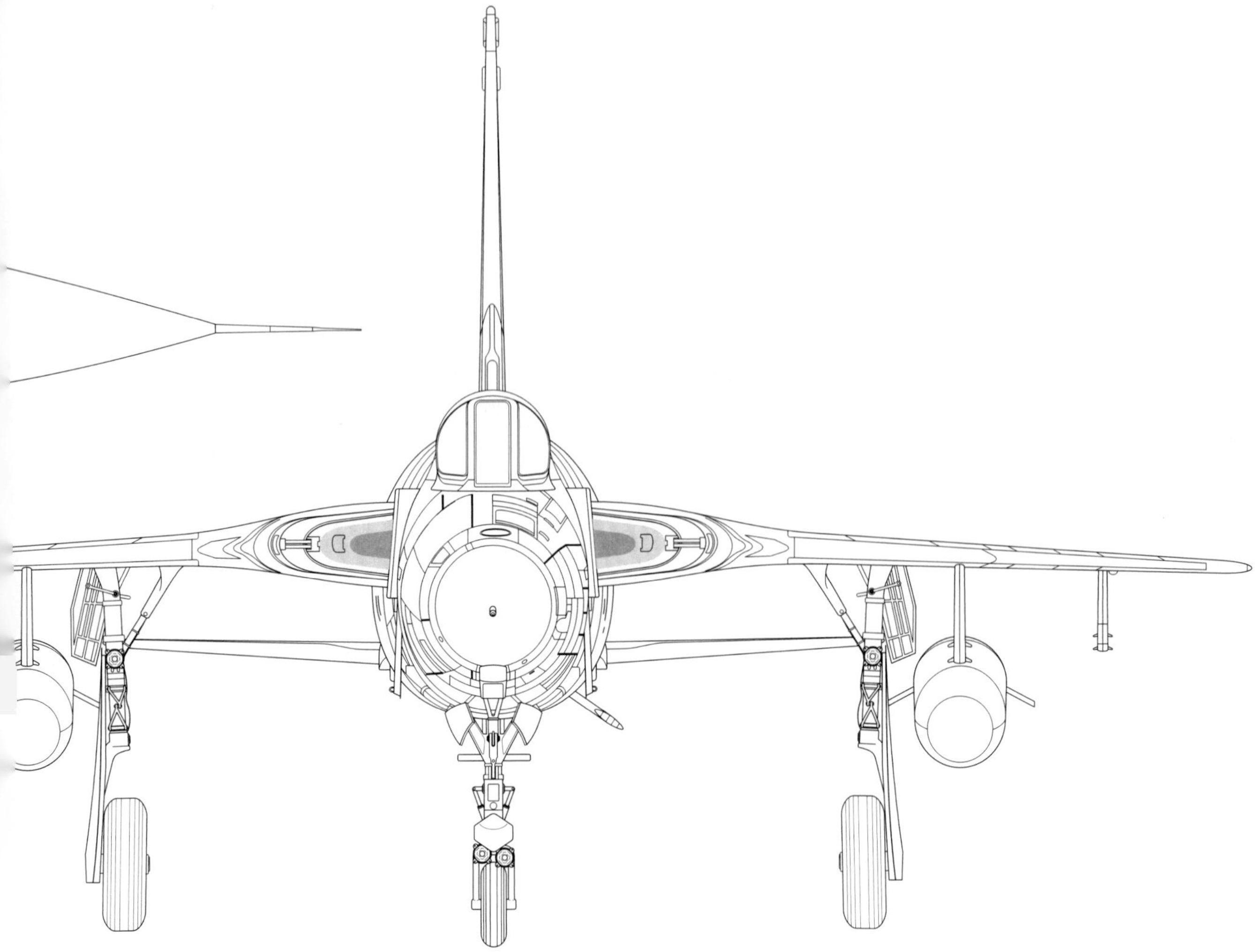

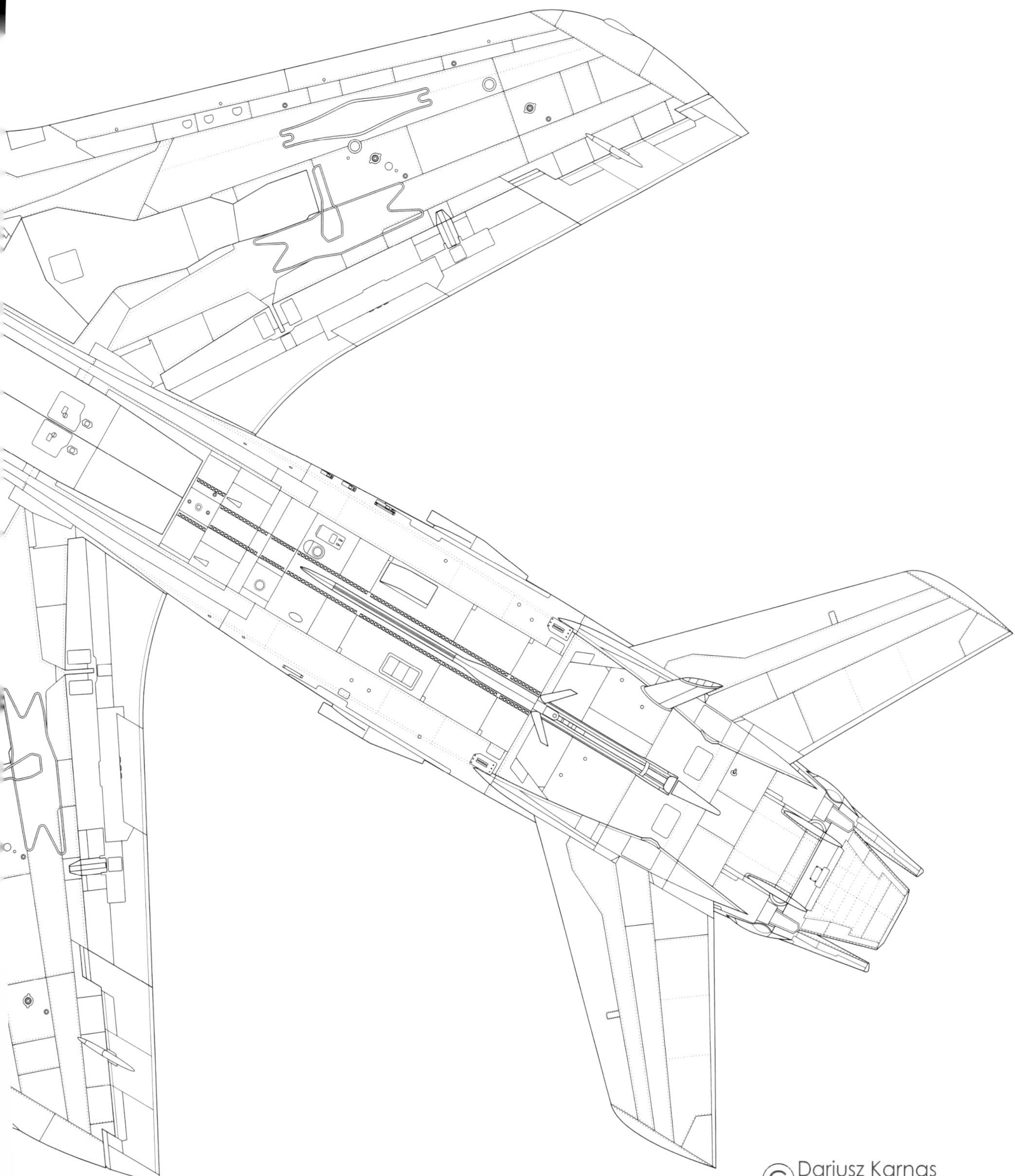

1/48 Scale Plans

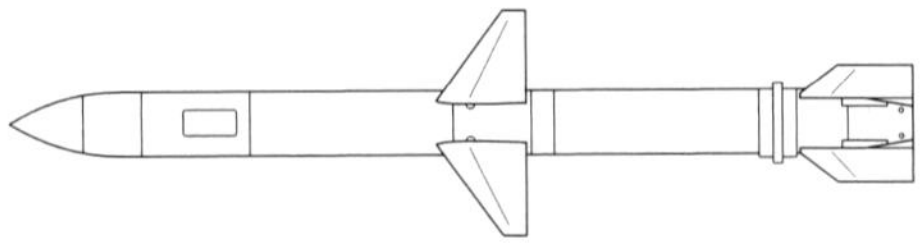

AGM-45

AIM-9 Sidewinder
a short-range air-to-air missile

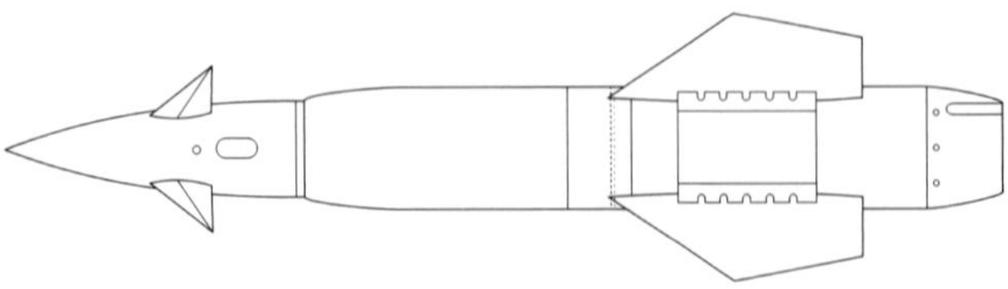

AGM-12

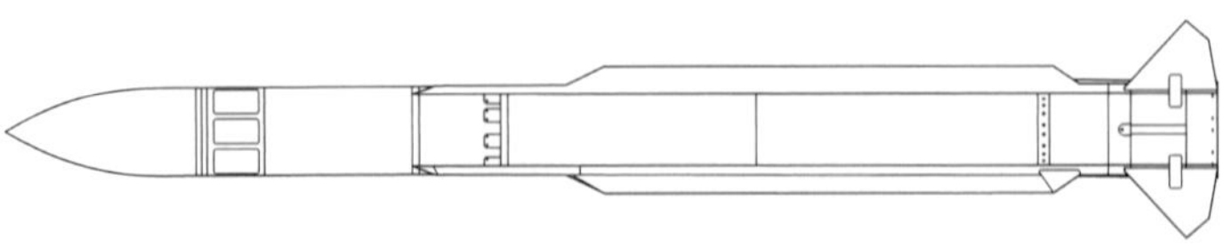

AGM-78

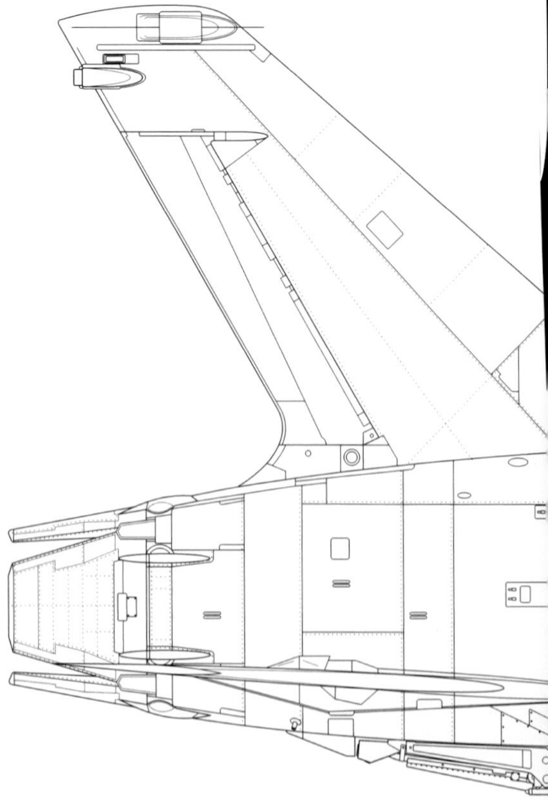

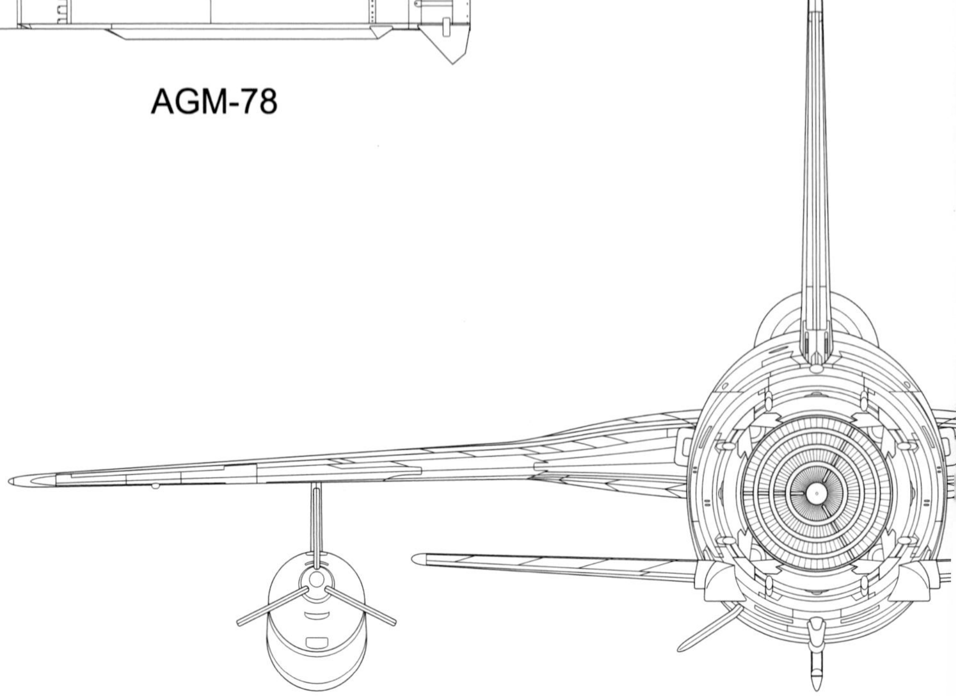

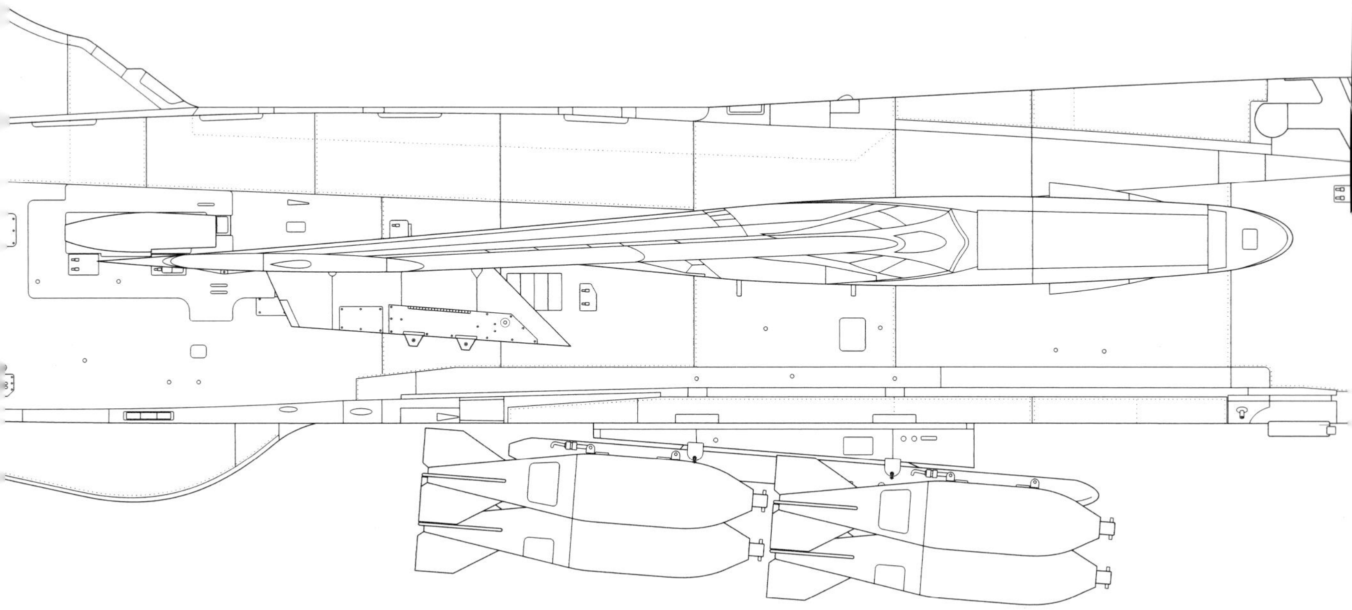

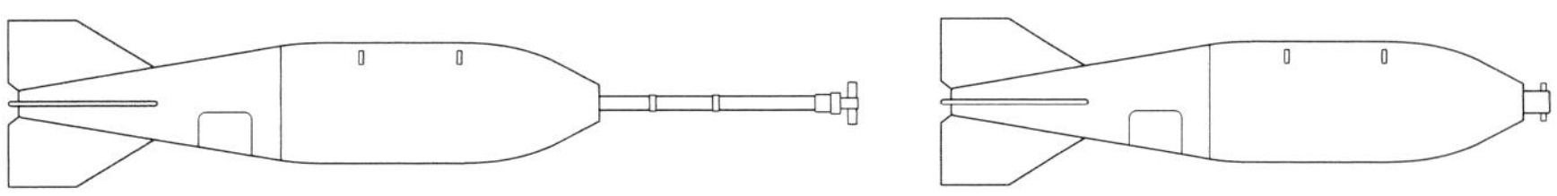

M 117 bombs (340 kg)

F-105D-31-RE

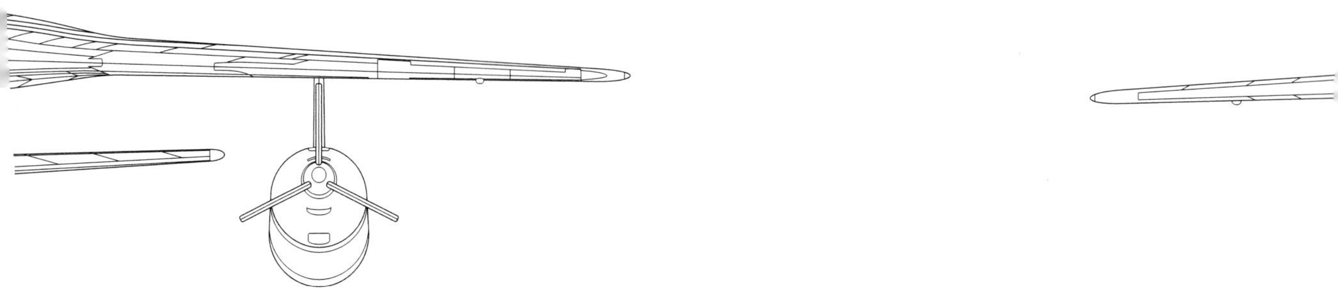

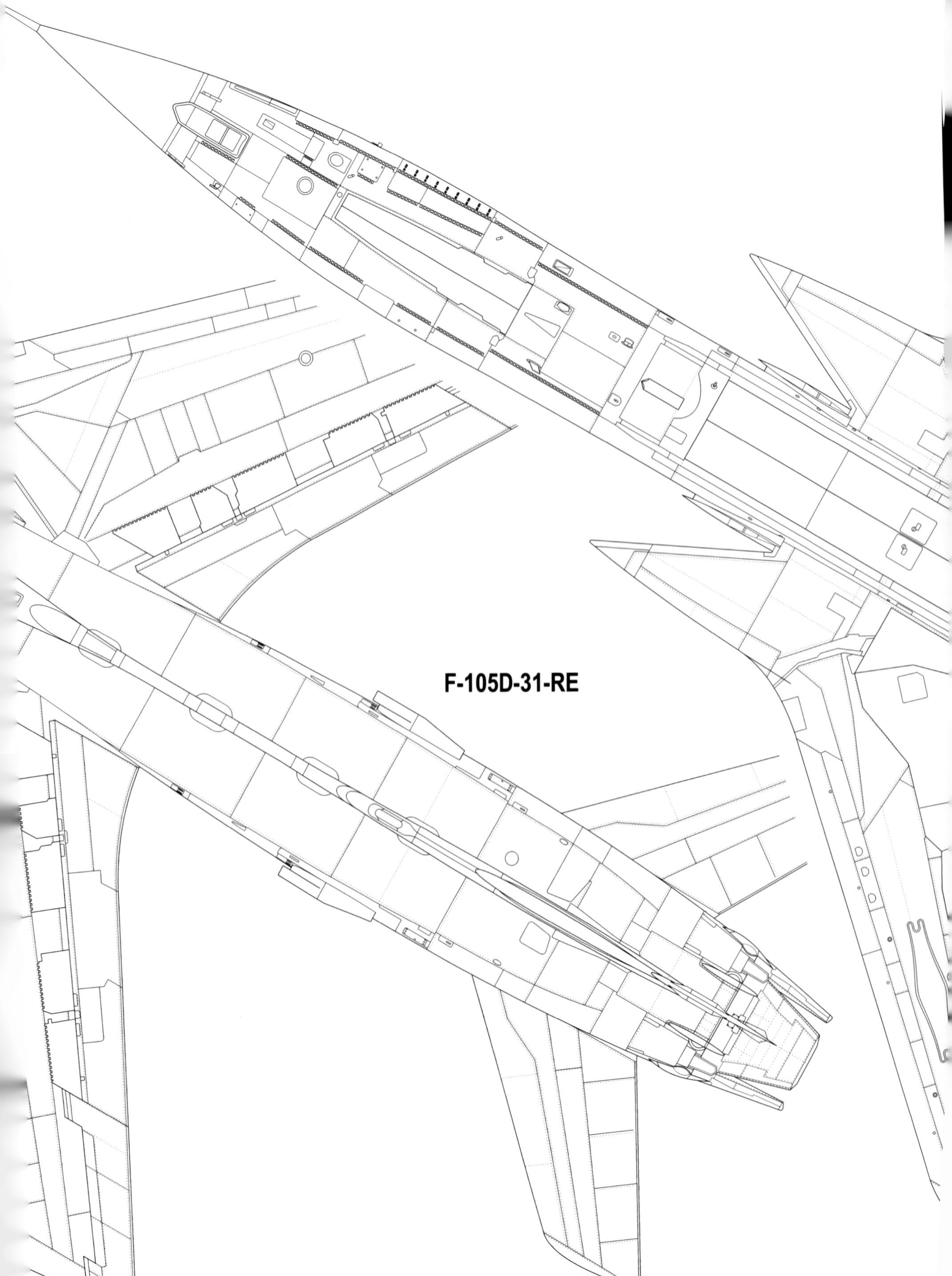

F-105D-31-RE

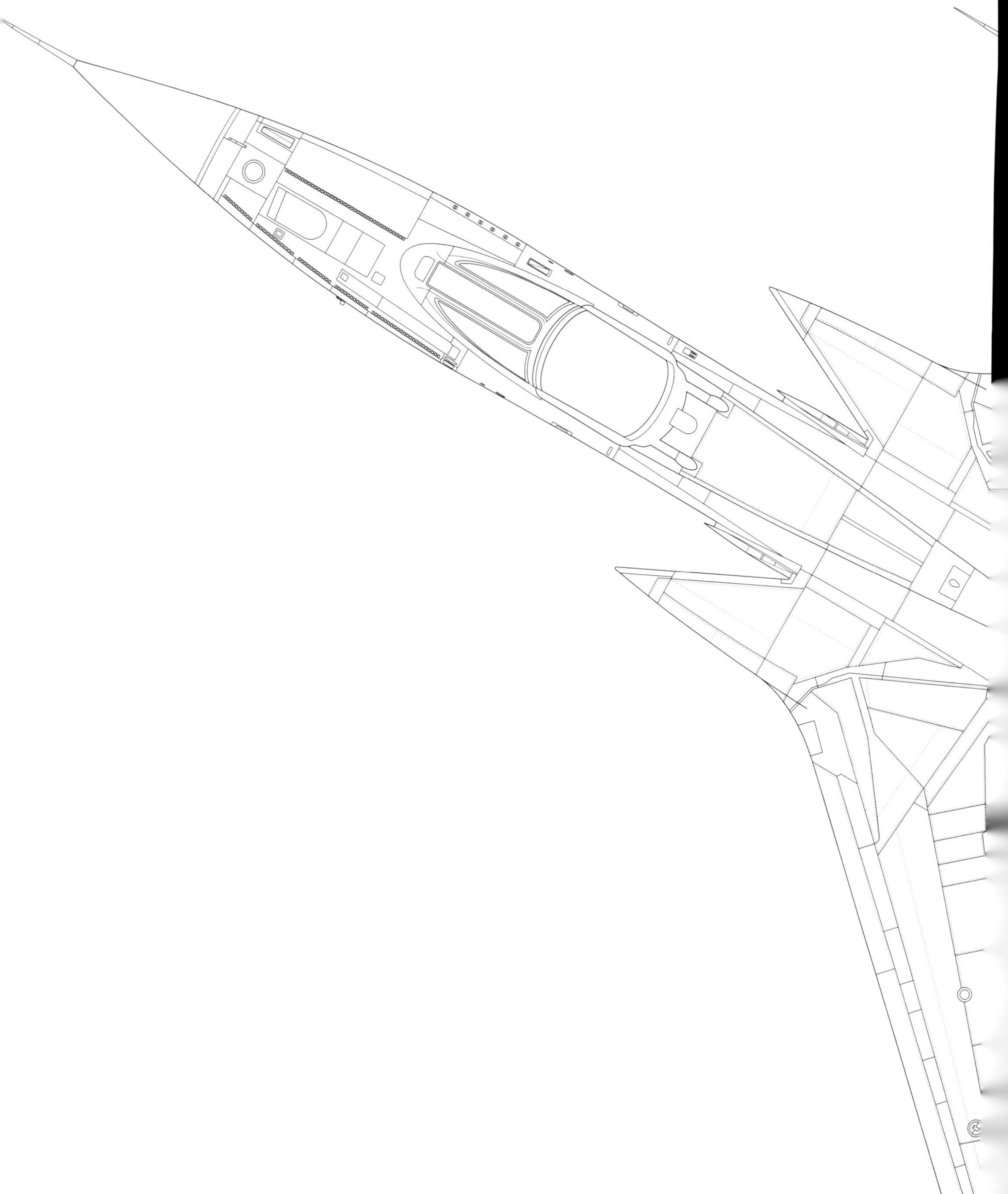

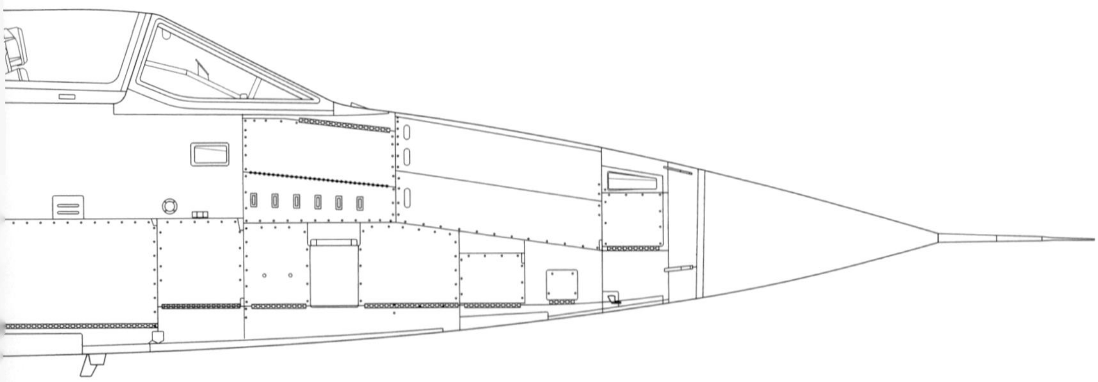

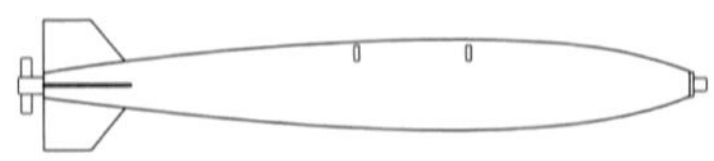

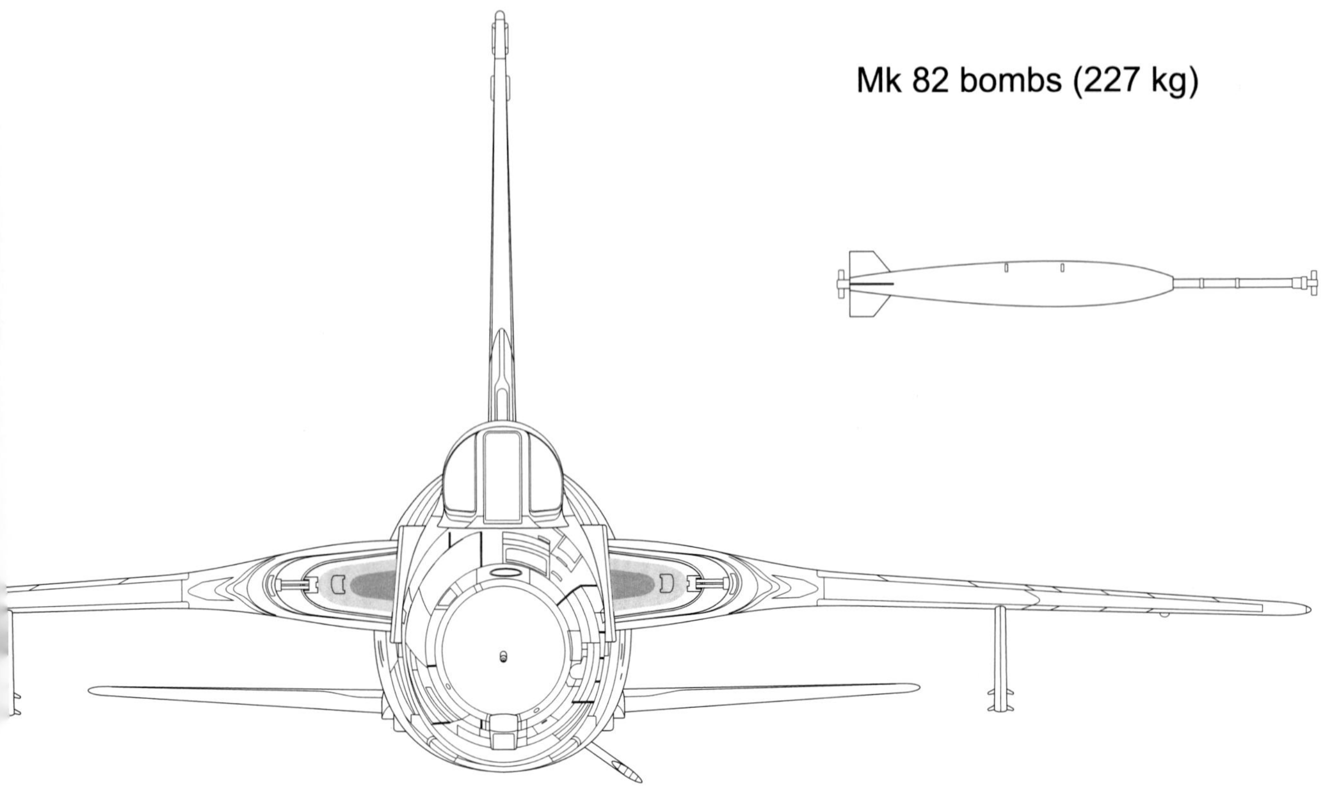

Mk 82 bombs (227 kg)

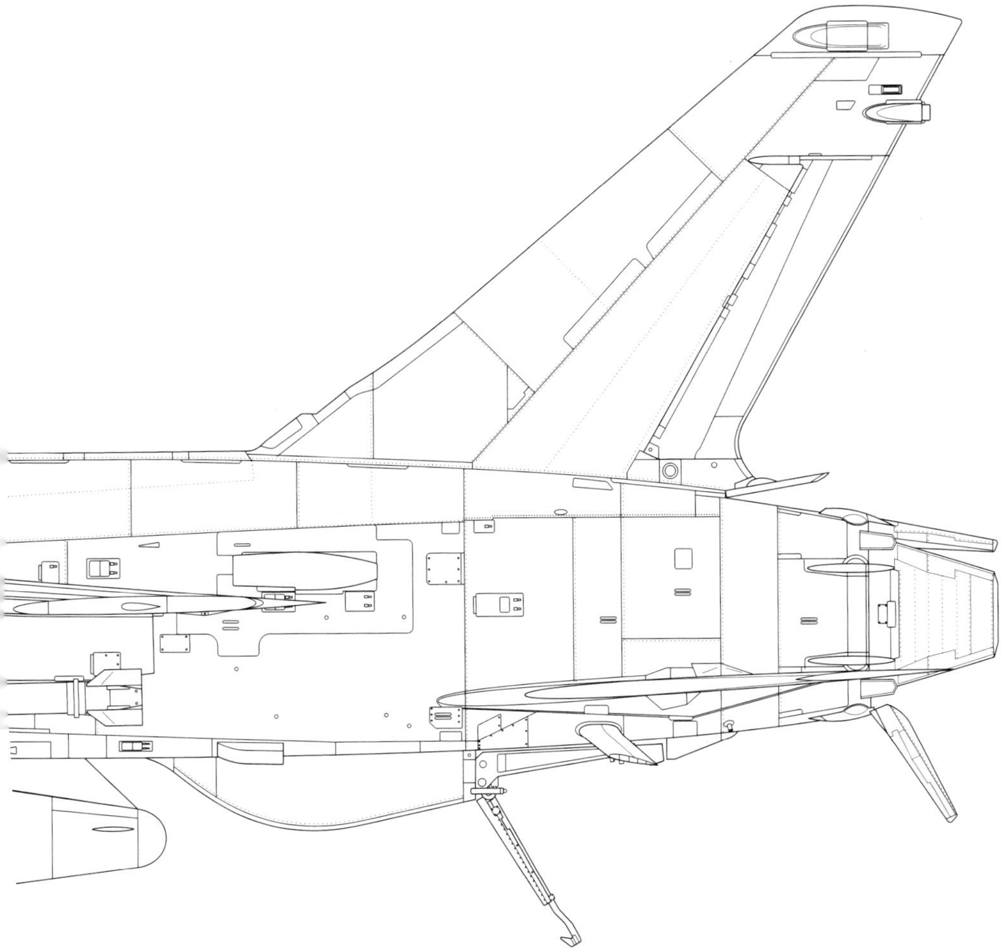

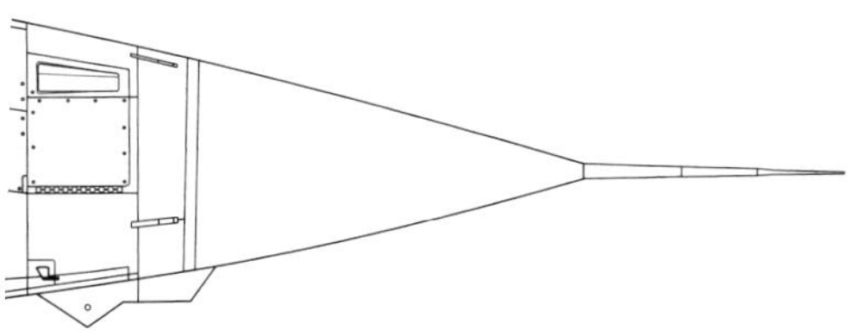

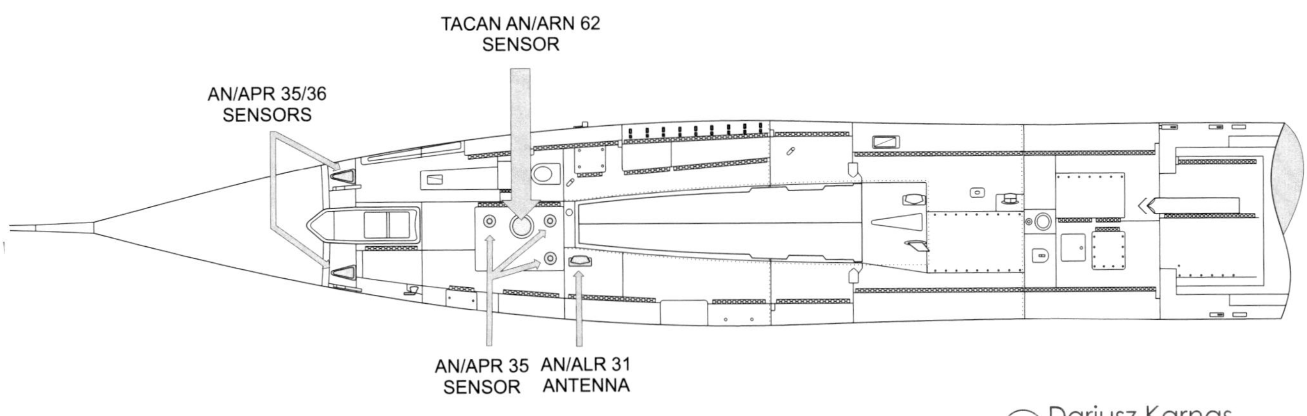
TACAN AN/ARN 62
SENSOR
AN/APR 35/36
SENSORS
AN/APR 35
SENSOR
AN/ALR 31
ANTENNA

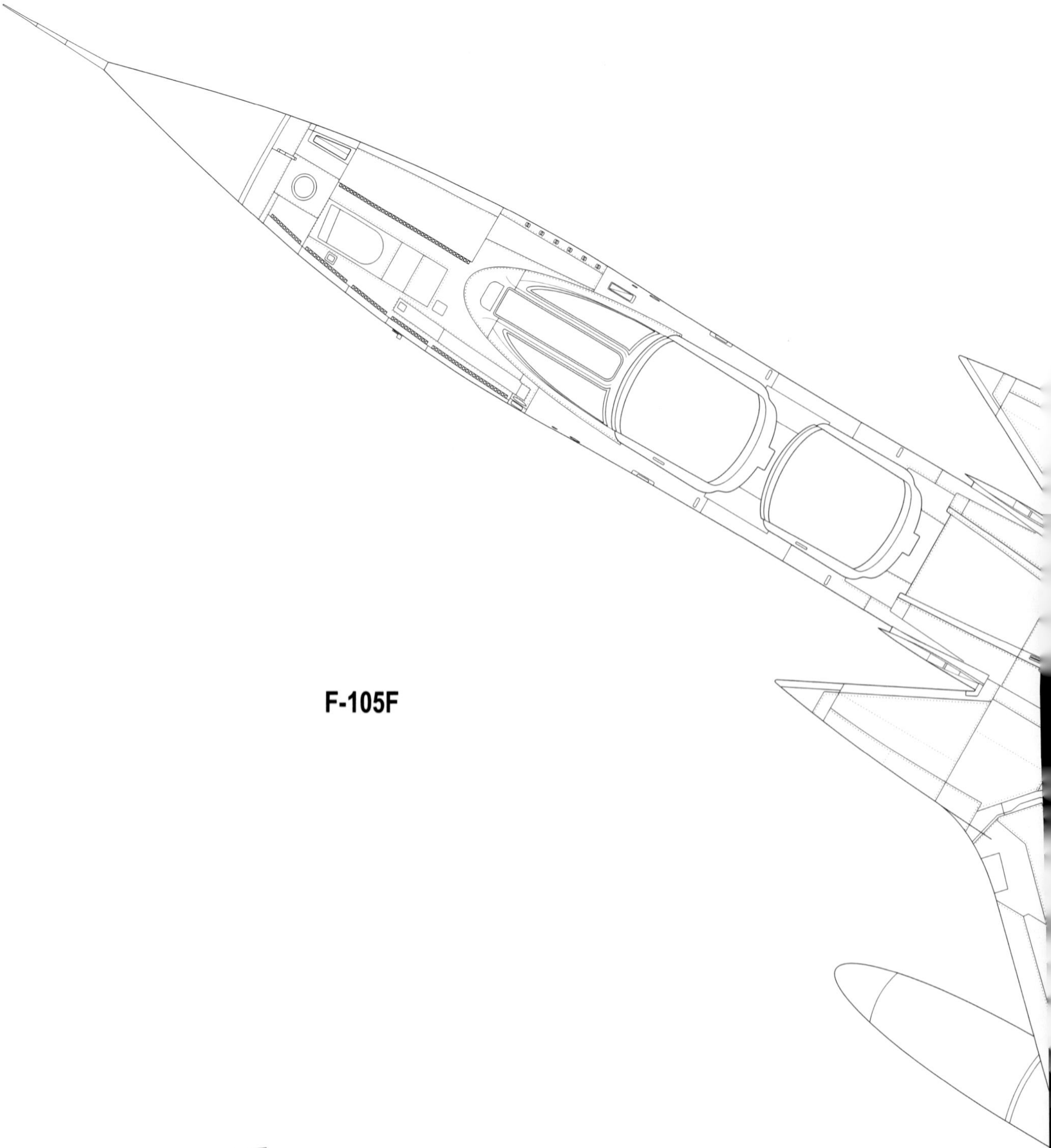

F-105F

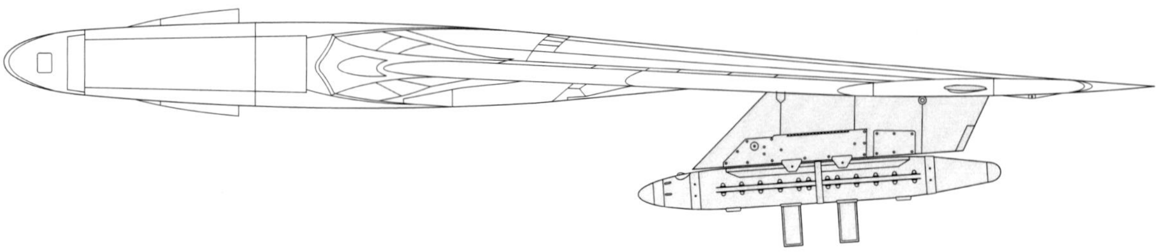

F-105G
wing with ALQ-87 on outer pylon

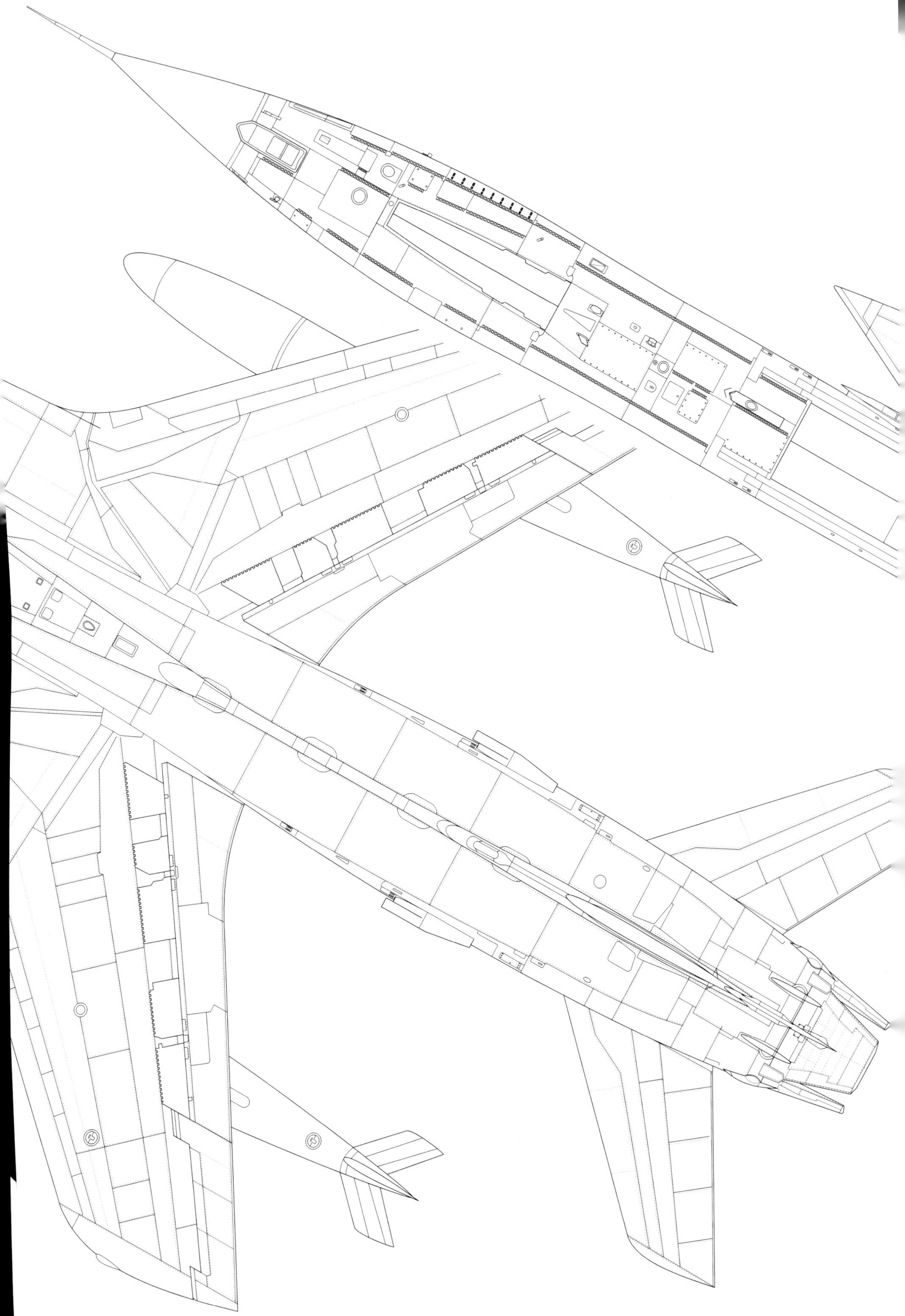

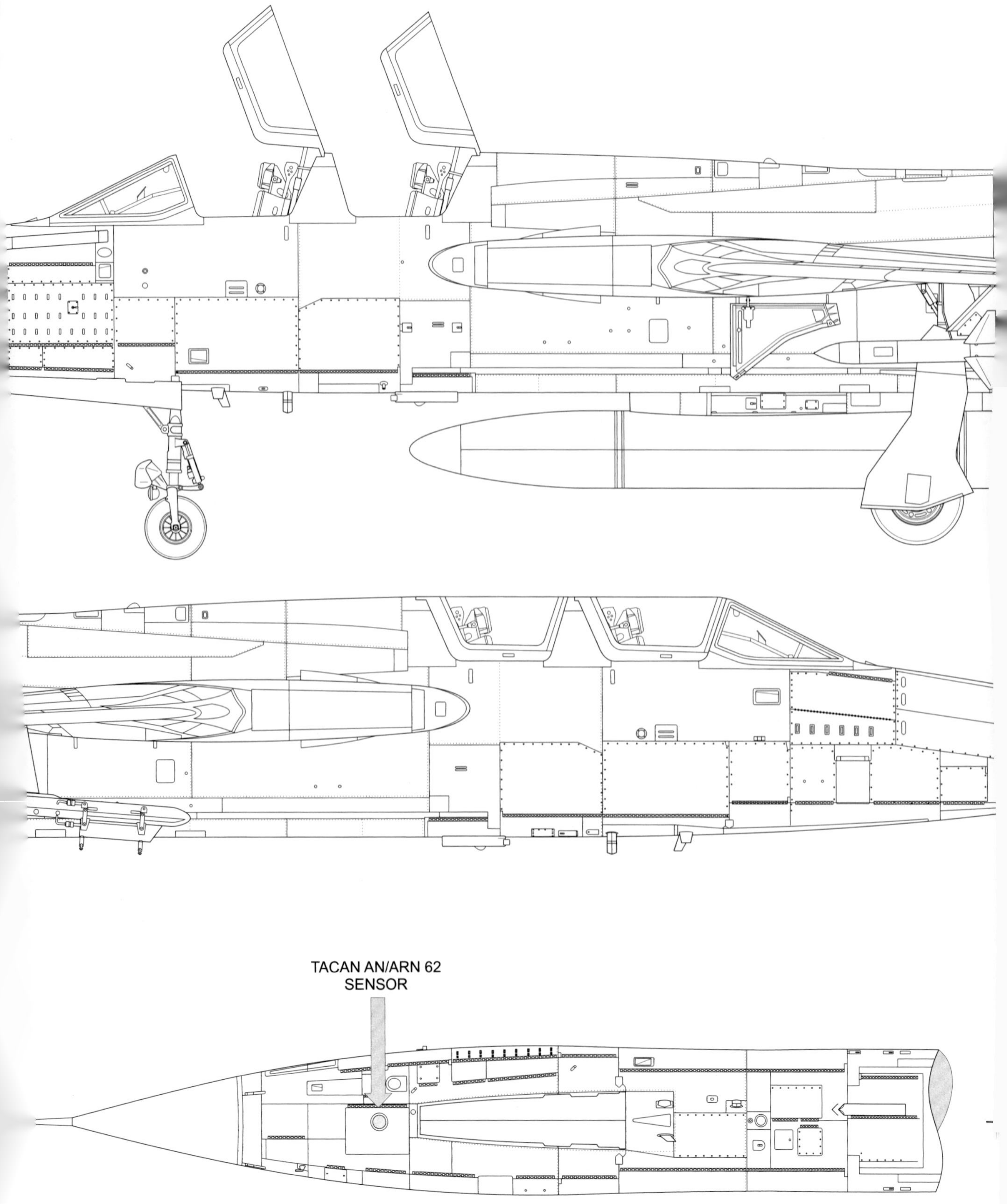

F-105F-1 RE

F-105F

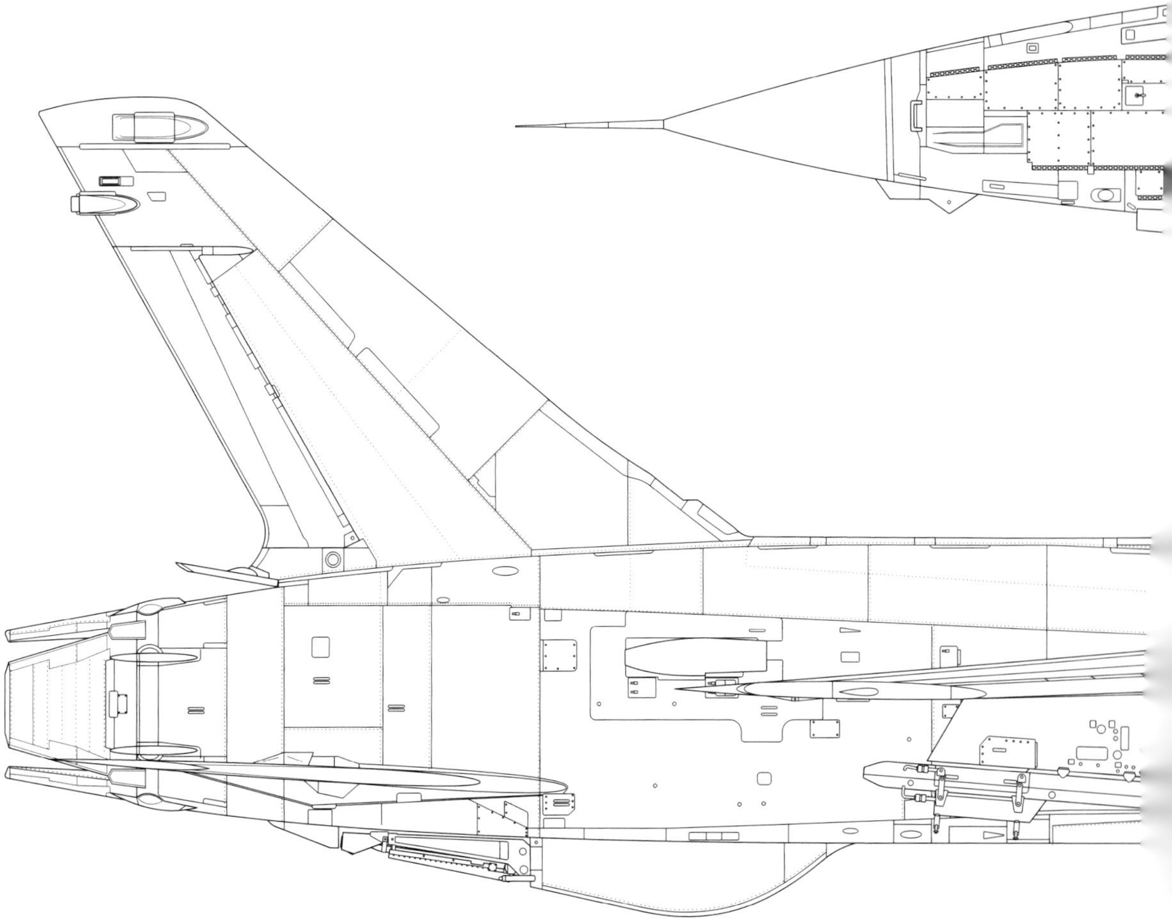

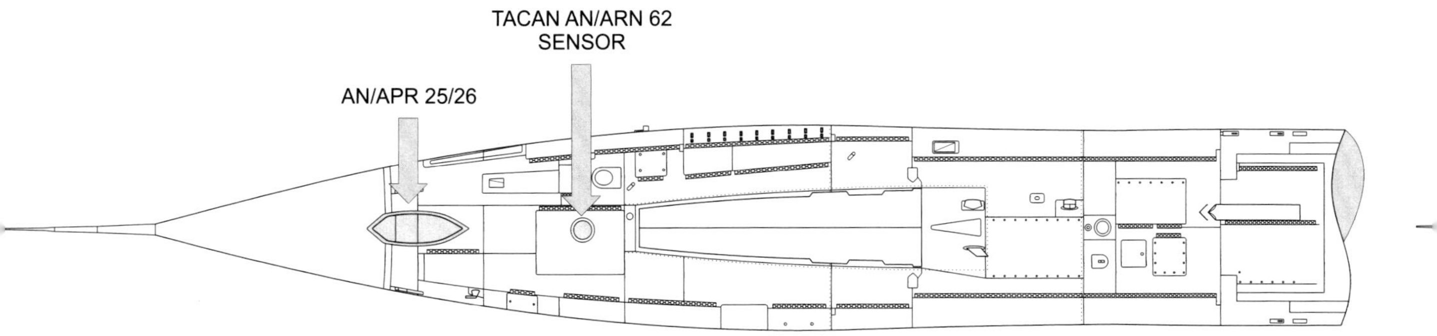

F-105F

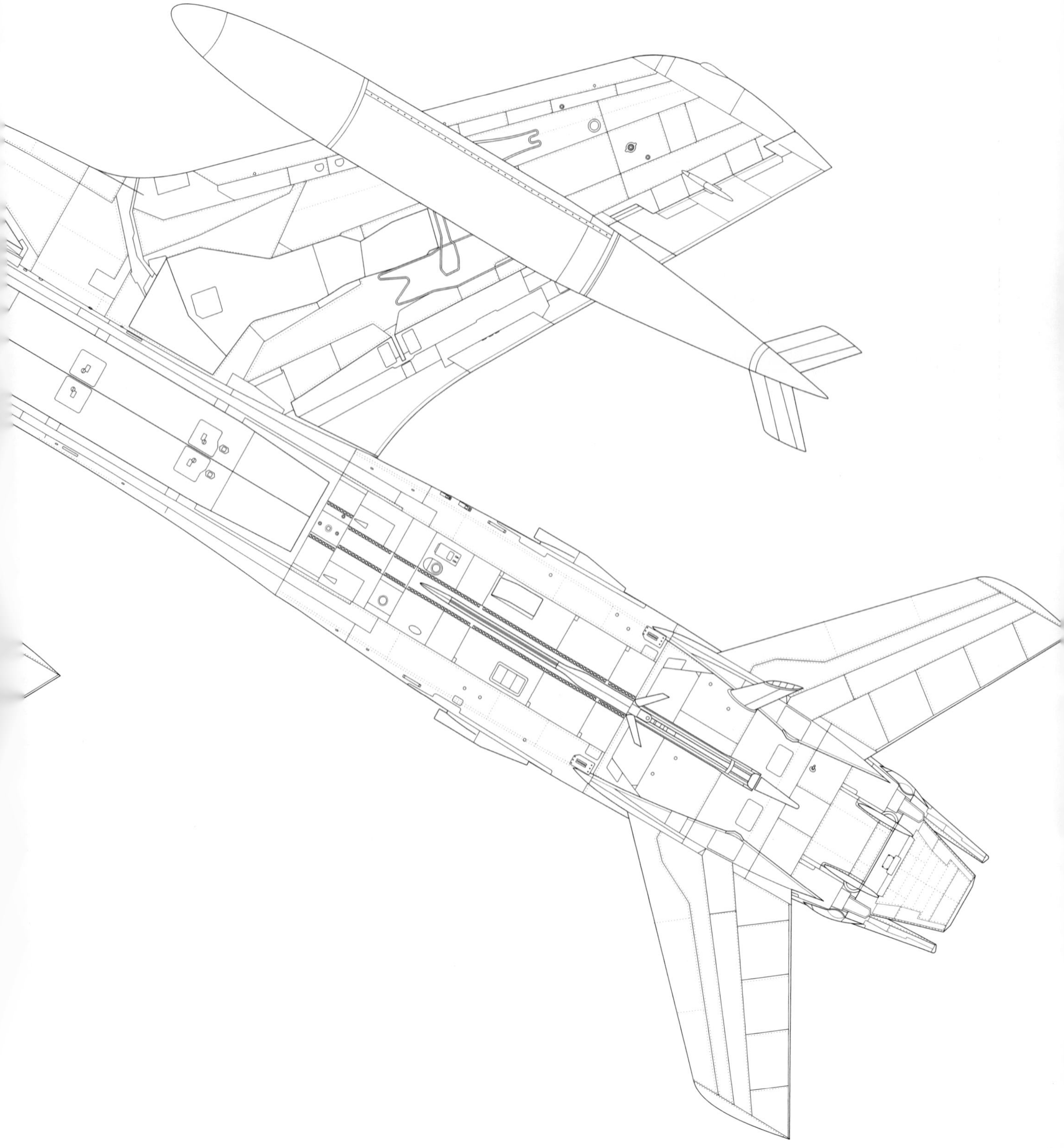